For Tala and Hugo

*Spark*an Interest: Scientists A to Z*

The images used in this book include public domain works and AI-generated illustrations created by the author with assistance from Flux AI Image Generator. All images have been cropped for inclusion.

For permission requests, write to the author at rachelreyesbrandwood@gmail.com.

NATIONAL LIBRARY OF AUSTRALIA

A catalogue record for this book is available from the National Library of Australia

ISBN: 978-1-7640267-0-3 (paperback)
ISBN: 978-1-7640267-1-0 (eBook)

Spark*an Interest

Scientists A to Z

Rachel Reyes-Brandwood

A is for Aristotle

Aristotle was a brilliant ancient Greek philosopher and scientist who studied nearly every subject, like biology, physics, logic, ethics, and politics. He helped start modern science by showing people how to observe the world, group things together, and think logically.

B is for Bohr, Neils

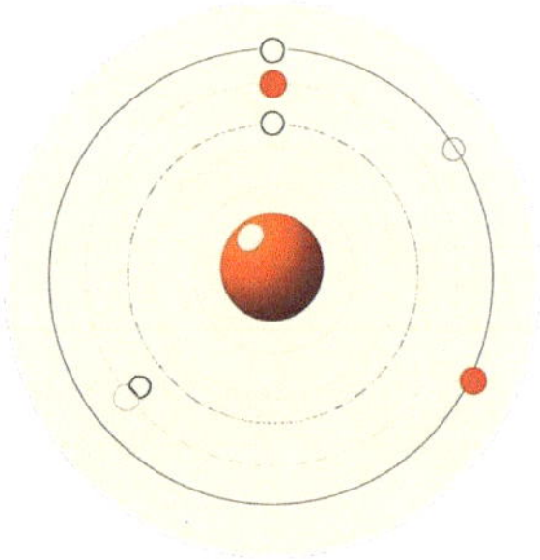

Neils Bohr was an ingenious scientist who figured out that tiny parts of atoms, called electrons, move around the centre in set paths, like planets around the sun. This helped us understand how everything in the world is made.

C is for Curie, Marie

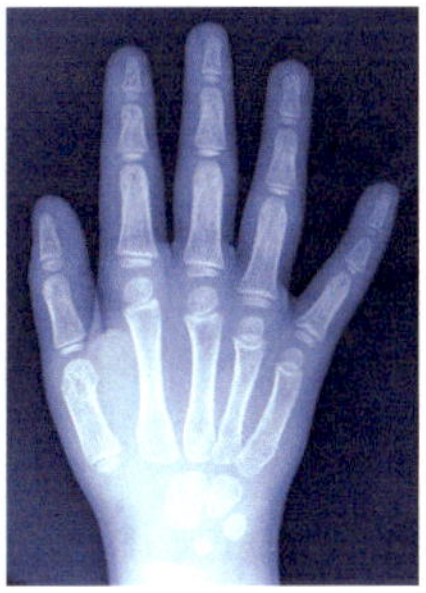

Marie Curie (born Maria Skłodowska) was an exceptional scientist who discovered two special elements, radium and polonium, and helped create X-ray machines that let doctors see inside the body. She won two big awards called Nobel Prizes for her amazing work.

D is for Darwin, Charles

Charles Darwin was an eminent scientist who discovered that animals and plants slowly change overtime to better survive in their environments. This idea is called evolution, which he explained in his famous book *On the Origins of Species*.

E is for Einstein, Albert

$$E = mc^2$$

Albert Einstein was a visionary scientist who figured out amazing ideas like how time and space work together. He also created the famous equation $E=mc^2$, which showed that tiny bits of matter can turn into a lot of energy.

F is for Faraday, Michael

Michael Faraday was an inventive scientist who discovered how to make electricity using magnets. This helped create things like electric motors and generators that power so much of what we use today.

G is for Galilei, Galileo

Galileo Galilei was a revolutionary scientist who built a powerful telescope to study the sky. With this, he discovered that moons orbit Jupiter, proving that the Earth is not the centre of everything but just part of something much bigger.

H is for Hawking, Stephen

Stephen Hawking was a brilliant scientist who studied the stars and black holes. Despite having a condition that limited his physical movement, he wrote books that made it easy for people to learn about the universe.

I is for Ibn al-Haytham

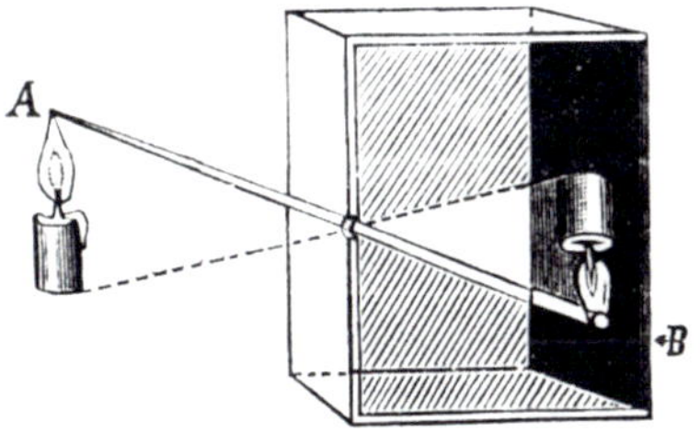

Ibn al-Haytham (aka Alhazen) was a pioneering scientist who discovered how light helps us see. He was one of the first people to understand how vision works which helped create things like cameras.

J is for Joule, James Prescott

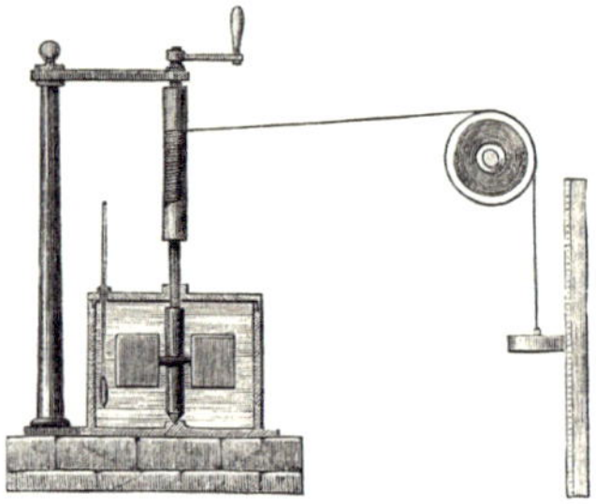

James Prescott Joule was an influential scientist who discovered that heat and energy are connected. The unit of energy is named after him because his work helped create the idea that energy can't be destroyed.

K is for Kepler, Johannes

Johannes Kepler was a meticulous scientist who found out that planets go around the Sun in a special oval paths called ellipses. He also discovered that planets move faster when they're close to the Sun and slower when they're far away, which helped us understand how our whole solar system works.

L is for Lavoisier, Antoine

Antoine-Laurent de Lavoisier was a foundational scientist who discovered oxygen and figured out that things don't disappear when they burn but just change form. He also helped create names for chemicals, which is why people call him the father of modern chemistry.

M is for Maxwell, James Clerk

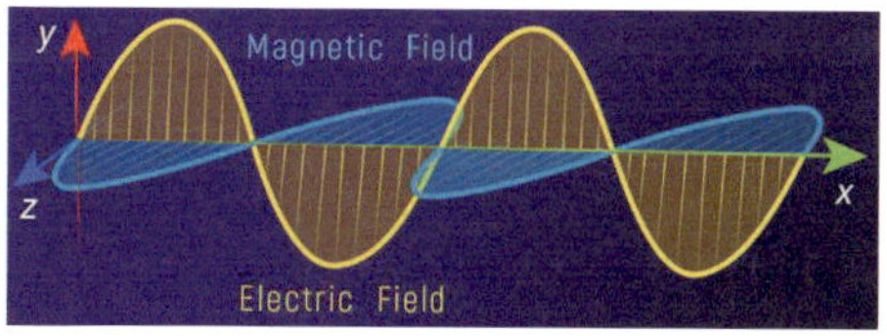

James Clerk Maxwell was an ingenious scientist who figured out that light, electricity, and magnets are all connected. He also discovered that they work like invisible waves called electromagnetic waves, which let us use radios and see all the colours of light around us.

N is for Newton, Isaac

Sir Isaac Newton was a groundbreaking scientist who discovered how gravity works, created the three laws of motion to explain how things move, and co-invented new maths called calculus. All these helped us understand how the universe works.

O is for Oppenheimer, J. Robert

J. Robert Oppenheimer was a controversial scientist who led the development of the first atomic bomb during World War II. He later expressed reservations about the consequences of creating this dangerous weapon.

P is for Pasteur, Louis

Louis Pasteur was an innovative scientist who discovered that tiny germs can make us sick. He then invented ways to make food safe, like heating milk to get rid of bad germs, so we don't get sick.

Q is for Quetelet, Adolphe

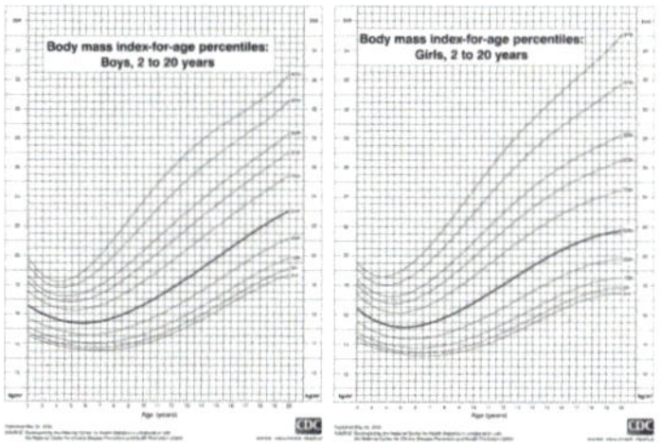

BMI charts for boys and girls

Lambert Adolphe Jacques Quetelet was a methodical scientist who used statistics or numbers to learn about people. He thought about what makes people typical in different ways and even helped create a tool that doctors use to understand how body size relates to health, which is now called the Body Mass Index (BMI).

R is for Rutherford, Ernest

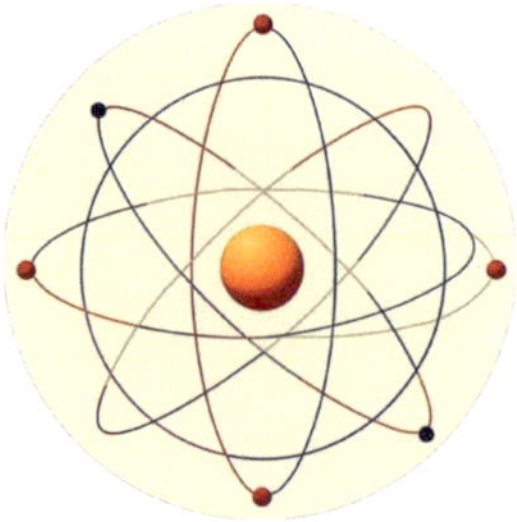

Ernest Rutherford was an insightful scientist who discovered that atoms have a tiny centre called a nucleus, like the Sun in the solar system. He also helped us understand radioactivity and how atoms can change.

S is for Schrödinger, Erwin

Erwin Schrödinger was an imaginative scientist who came up with tricky ideas about how tiny things like electrons move. He used a famous thought experiment with a cat in a box to explain how strange and uncertain the world of atoms can be.

T is for Tesla, Nikola

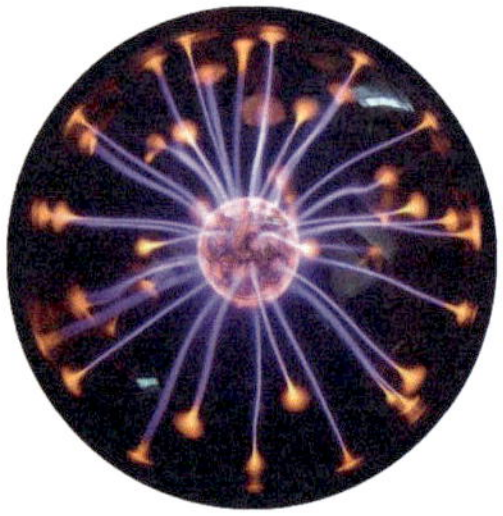

Nikola Tesla was a brilliant inventor who created a special kind of electricity called alternating current (AC) that powers our homes today. He also invented the Tesla coil which produces strong electricity and big sparks that were used in early radios and is still used in entertainment, science shows and in finding tiny holes.

U is for Urey, Harold

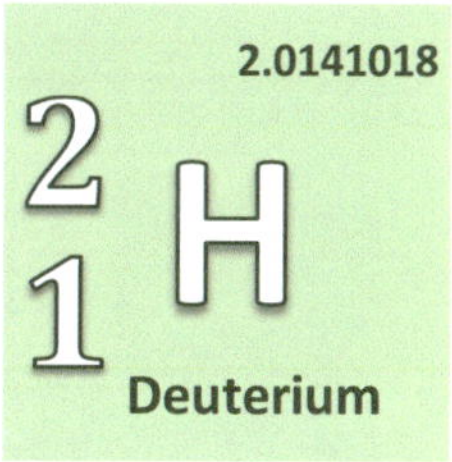

Harold Urey was an exploratory scientist who discovered deuterium, a special kind of hydrogen that helps make electricity and is used as a tracer to study how plants grow and how our bodies work. He also explored how life might have started on Earth and what the planets are made of.

V is for Virchow, Rudolf

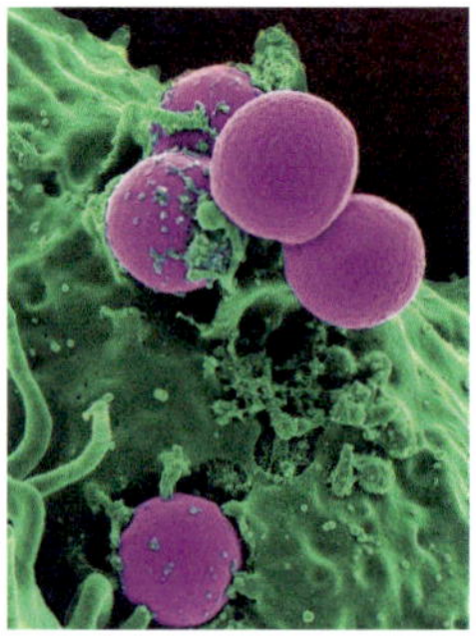

Rudolf Virchow was a perceptive doctor who figured out that sicknesses often start from tiny problems inside our body's cells. He also helped us understand that all cells come from other cells, just like how baby animals come from their parents.

W is for Watson, James D.

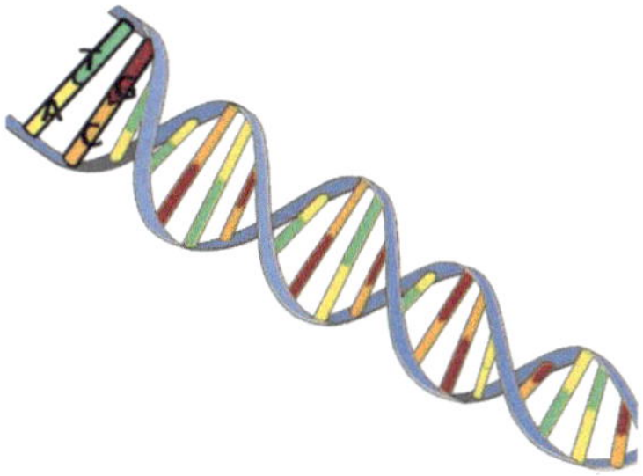

James Dewey Watson is a groundbreaking scientist who helped find out that DNA, the tiny instructions inside all living things, looks like a twisted ladder. This discovery helped us understand how we grow and why we look the way we do, because DNA contains all the information that makes us who we are.

X is for Xing, Yi

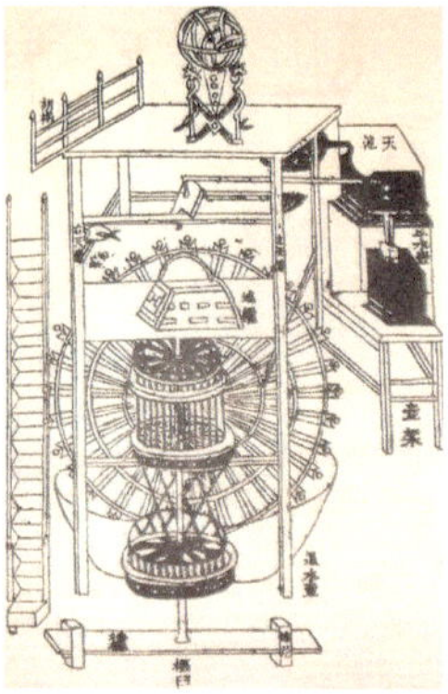

Yi Xing was clever monk and scientist from ancient China who invented the first mechanical clock powered by water. He also studied the stars to help make better calendars.

Y is for Yang, Chen-Ning

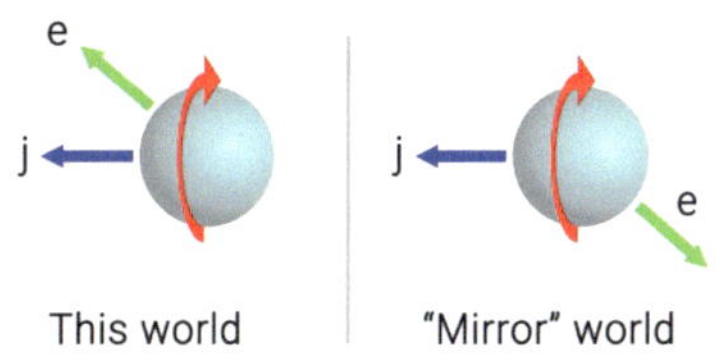

Chen-Ning Yang was a brilliant physicist who co-discovered that tiny particles don't always behave the same way when you flip them like a mirror, an idea called parity nonconservation. He won a big award called the Nobel Prize for helping us understand how these tiny particles work together.

Z is for Zwicky, Fritz

Fritz Zwicky was a visionary scientist who studied space and discovered important things like dark matter, supernovas, and how galaxies can bend light. He showed us that space is full of wonder and exciting things to discover.

Description of Featured Images

A: Cover of Aristotle's collection of six works on logic with the Latin title, *Aristotelis Stagiritae Peripateticorum Principis Organum*, which roughly translates to "The Instrument of Aristotle from Stagira, the Leader of Peripatetics (Aristotle's followers)"

B: Stylised representation of Bohr's atomic model where the electrons orbit around the nucleus in circular paths at specific fixed distances corresponding to discrete energy levels

C: Coloured X-ray image of a 5-year old boy's hand

D: Depiction of human evolution from a common ancestor with primates

E: Einstein's famous physics equation which states energy (E) is equal to mass (m) multiplied by the speed of light (c) squared

F: Model of Faraday's invention called Faraday disk/Faraday disk generator, an early electromagnetic generator which converts mechanical energy to electrical energy

G: Stylised image of a telescope pointed to planet Jupiter, symbolising Galileo's groundbreaking observation, which supported the Copernican heliocentric model and challenged the then prevailing geocentric view of the universe

H: Photo of a black hole consuming a star, technically referred to as a tidal disruption event (TDE)

I: Diagram of a camera obscura, an optical device that projects an inverted image of the external word onto a surface inside a darkened space

J: Drawing of Joule's apparatus, a device designed by Joule to measure the mechanical equivalent of heat, which demonstrated that mechanical energy can be converted to heat energy and contributed to the development of the principle of energy conservation

K: Stylised representation of the solar system, featuring the Sun and its nine planets: Mercury, Venus, Earth, Mars, Jupiter, Saturn, Uranus, Neptune, and Pluto, in their respective positions and orbits around the Sun

L: Stylised illustration of Lavoisier's experiment involving the decomposition of mercury oxide, a discovery that led to the identification of oxygen as an element

M: Graphical representation of electromagnetic waves, illustrating the synchronised oscillations of electric and magnetic fields as they propagate through space

N: Stylised image of an apple falling from a tree, reminiscent of the moment that inspired Newton to formulate the law of universal gravitation

O: Cropped photograph of a mushroom cloud resulting from nuclear bomb testing

P: Stylised image of a carton of milk labelled “Pasteurised”

Q: Charts of body mass index-for-age percentiles, for boys and girls, respectively, for ages 2 to 20 years

R: Stylised representation of Rutherford’s atomic model, featuring a dense, positively charged nucleus at the centre, surrounded by electron in circular orbits, with most of the atom being empty space

S: Stylised illustration of Schrödinger’s cat thought experiment, where a cat in a sealed box is simultaneously alive and dead until observed, highlighting the paradox of quantum superposition

T: Cropped photograph of a plasma lamp, based on Tesla’s pioneering work with high-voltage currents, featuring colourful electrical discharges that respond to touch within a sphere filled with inert gases like neon and argon

U: Hypothetical periodic table entry for Deuterium, an isotope of hydrogen, with an atomic mass of 2.0141017926 Da, atomic number of 1, with one proton, one neutron and one electron

V: Microscopic image of white blood cells, which Virchow studied in his work on cellular pathology

W: Stylised illustration of the double helix structure of DNA (deoxyribonucleic acid), representing the twisted ladder-like arrangement of nucleotides that encode genetic information

X: An image from Su Song’s book depicting his astronomical clock tower, which incorporated the escapement mechanism developed by Yi Xing in the 8th century, an innovation crucial for the precision and automation of astronomical timekeeping in ancient China

Y: Illustration depicting parity transformation, converting a physical system into its mirror image, which reveals the fundamental asymmetry discovered in weak interactions, where parity is not conserved

Z: Image of shock wave around a supernova captured by the Hubble Space Telescope

Made in the USA
Las Vegas, NV
09 November 2026

57757236R00021